BoE 4035

HANS-GÜNTER HEUMANN

HITS FOR KIDS

DIE FETZIGSTEN ROCK- UND POPSTÜCKE ALLER ZEITEN IN LEICHTEN ARRANGEMENTS FÜR KLAVIER / KEYBOARD

THE MOST BRILLIANT ROCK AND POP SONGS EVER WRITTEN IN EASY ARRANGEMENTS FOR PIANO / KEYBOARD

①

INHALT / CONTENTS

		Seite/Page
①	With a Little Help from My Friends (John Lennon/Paul McCartney)	2
②	One Moment in Time (Albert Hammond/John Bettis)	4
③	Ice in the Sunshine (Holger-Julian Kopp/Hanno Haders)	6
④	Rock on the Rocks (Hans-Günter Heumann)	8
⑤	Still Loving You (Rudolf Schenker/Klaus Meine)	10
⑥	That's My Sound (Hans-Günter Heumann)	14
⑦	Oh, Pretty Woman (Roy Orbison/Bill Dees)	16
⑧	Midnight Special (Trad./ Hans-Günter Heumann)	19
⑨	Rock My Soul (Trad./ Hans-Günter Heumann)	20
⑩	Hit Boogie (Hans-Günter Heumann)	22

© Copyright MCMXC by Bosworth & Co., Berlin

BOSWORTH EDITION

With a Little Help from My Friends
(BEATLES)

Words & Music by John Lennon and Paul McCartney
Arr.: Hans-Günter Heumann

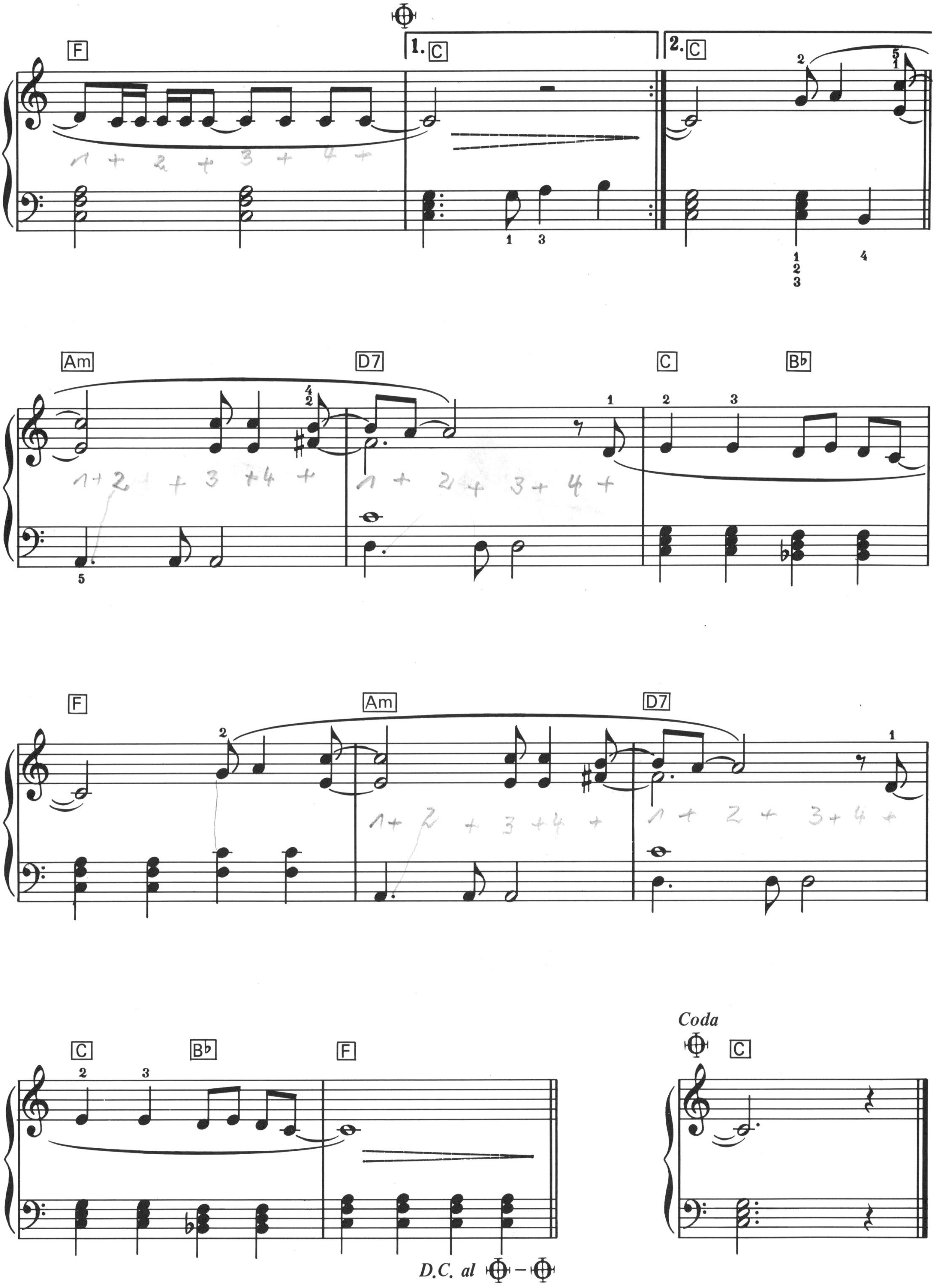
F
1. C
2. C
Am
D7
C
B♭
F
Am
D7
C
B♭
F
Coda
C
D.C. al

One Moment in Time
(WHITNEY HOUSTON)

Words and Music by Albert Hammond/ John Bettis
Arr.: Hans-Günter Heumann

Andante M.M. ♩ = 72-80

© 1988 Albert Hammond Music and John Bettis Music, USA.
Warner/Chappell North America, London/ P And P Songs Limited.
Reproduced by permission of Faber Music Limited.
All Rights Reserved. International Copyright Secured.

B. & Co. 24 928

Alle Rechte vorbehalten
All rights reserved

C
Am
Dm
G
f
F
G
C
Am
Dm
G
F
G
C
Am
Dm7
G
C
F
G7
D.S al
Coda
Dm7
Gsus4
G
C
Am
F
G
C
ff
rit.

Ice in the Sunshine

(BEAGLE MUSIC LTD.)

Words and Music by Holger-Julian Kopp / Hanno Harders
Arr.: Hans-Günter Heumann

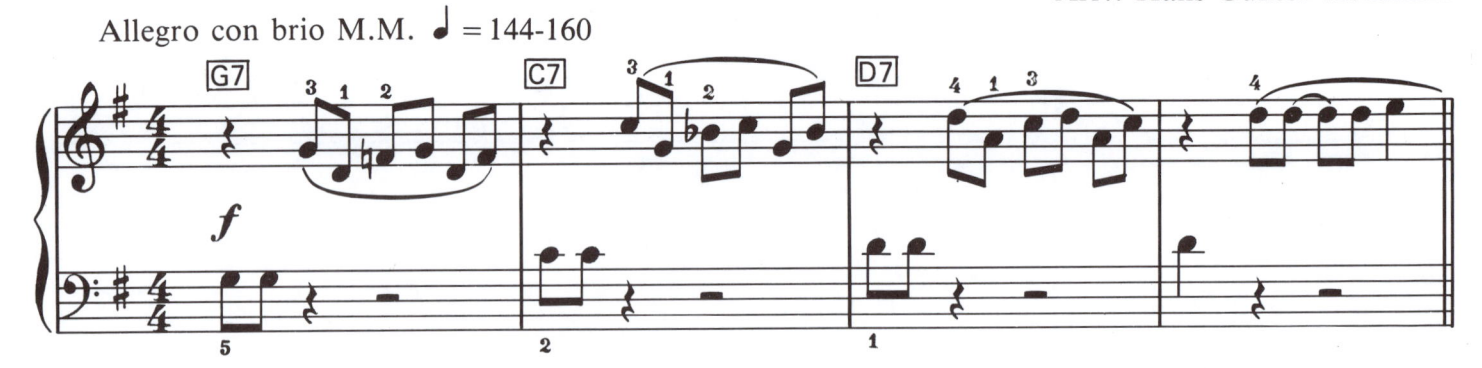

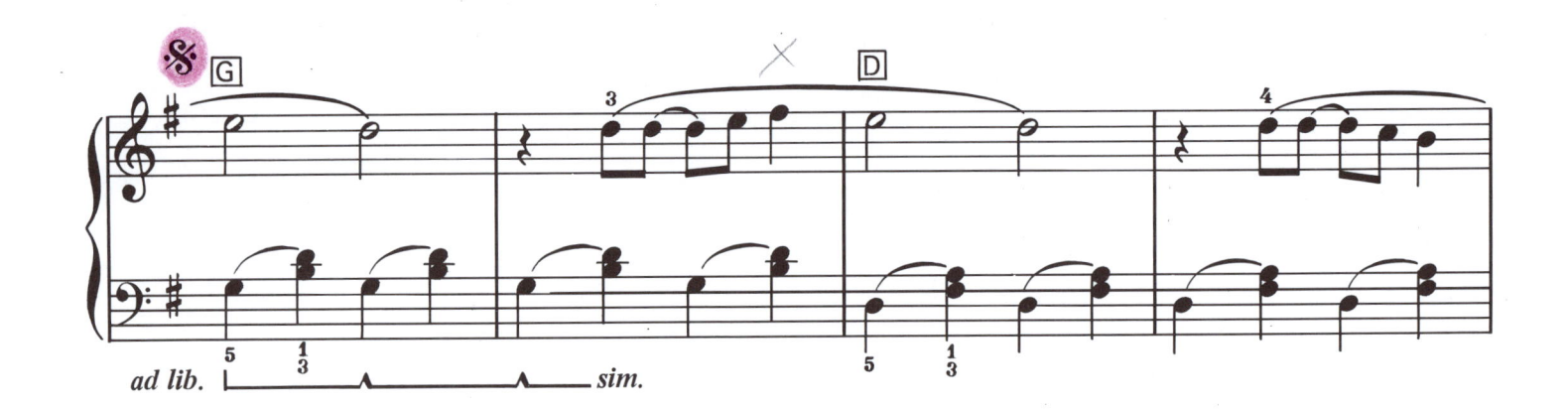

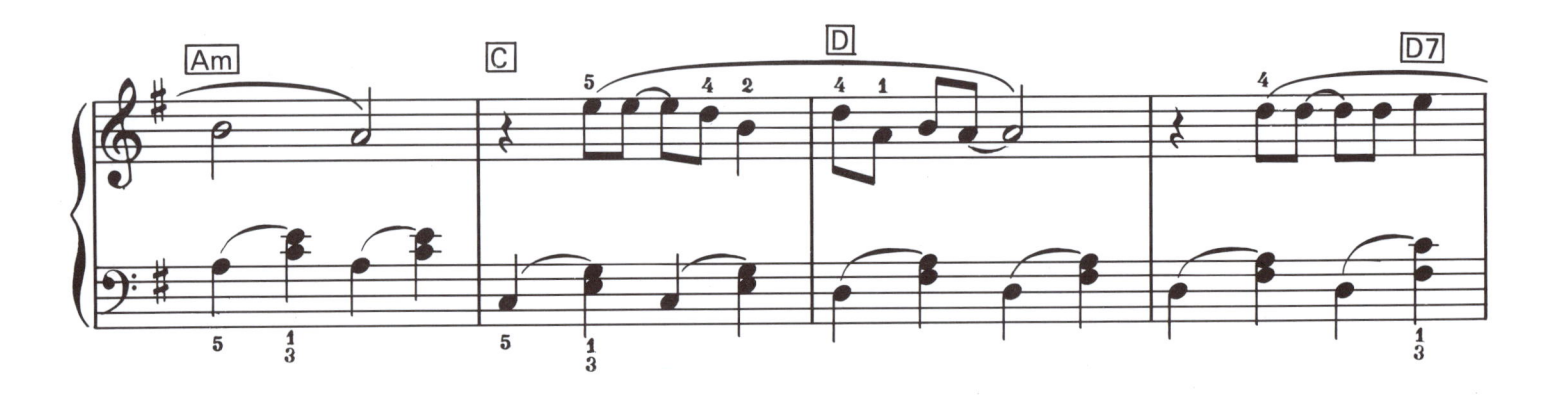

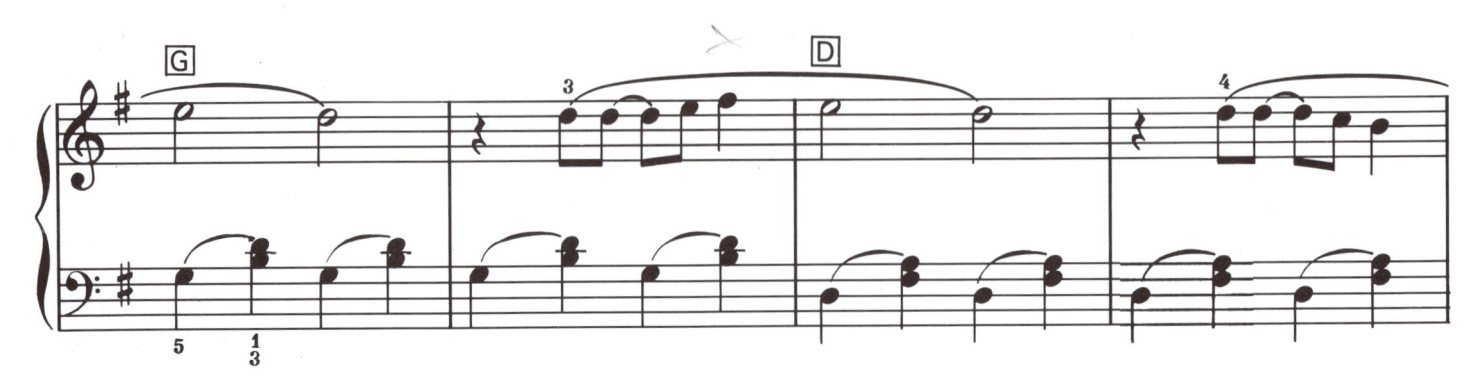

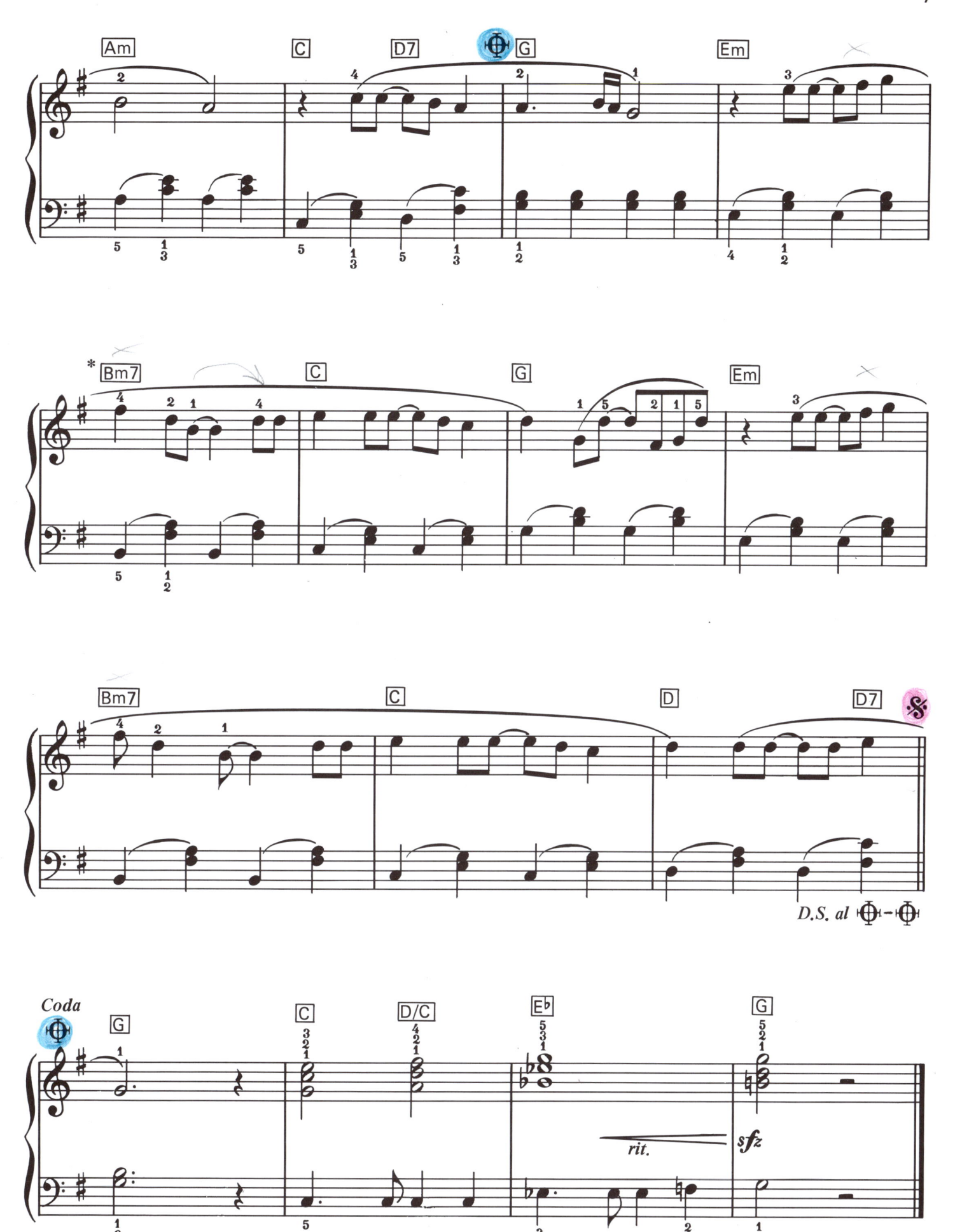

*Anmerkung für den Lehrer: Um Mißverständnisse zu vermeiden - die internationale Schreibweise für den Ton H ist B.

Rock on the Rocks

Music by Hans-Günter Heumann

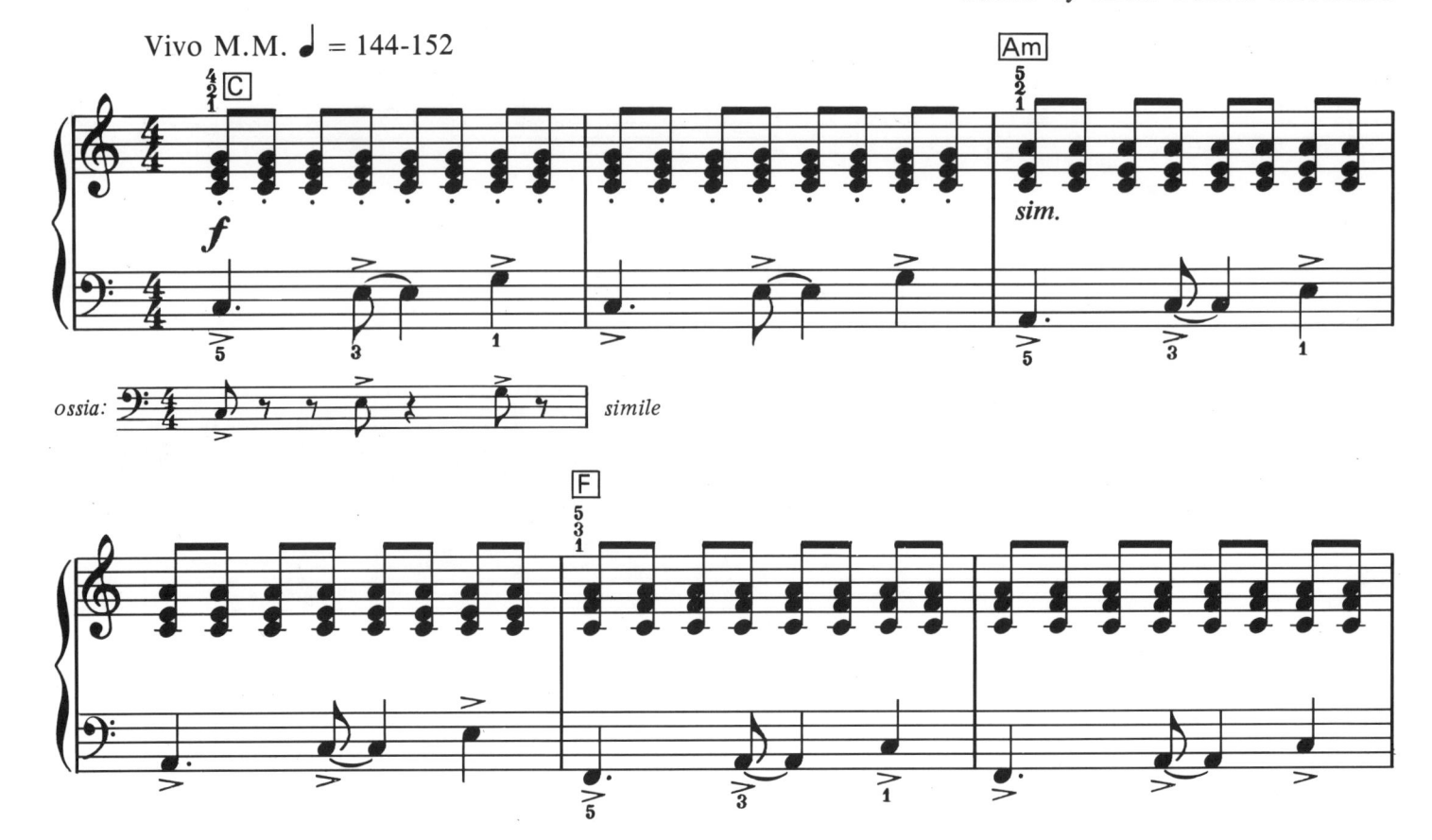

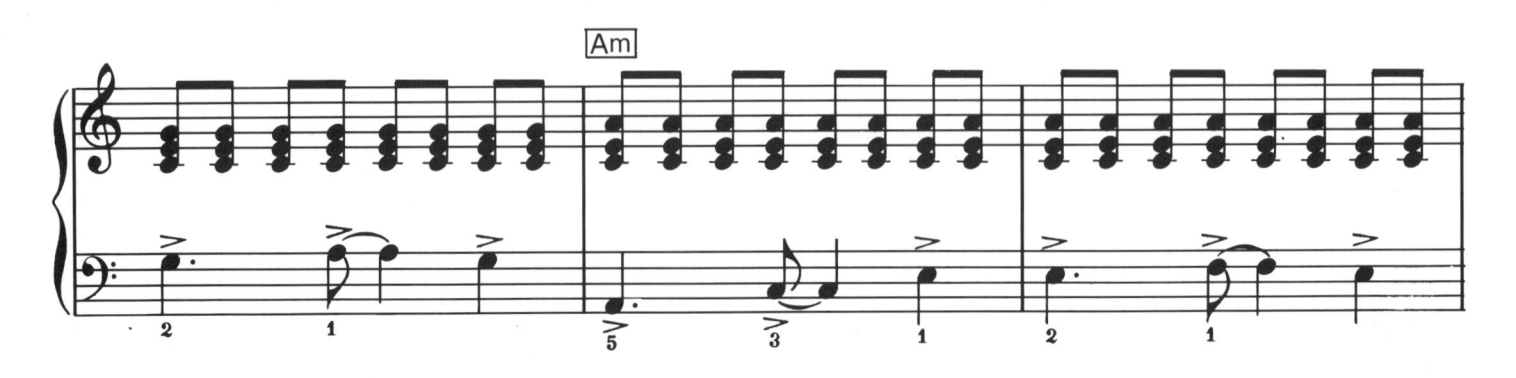

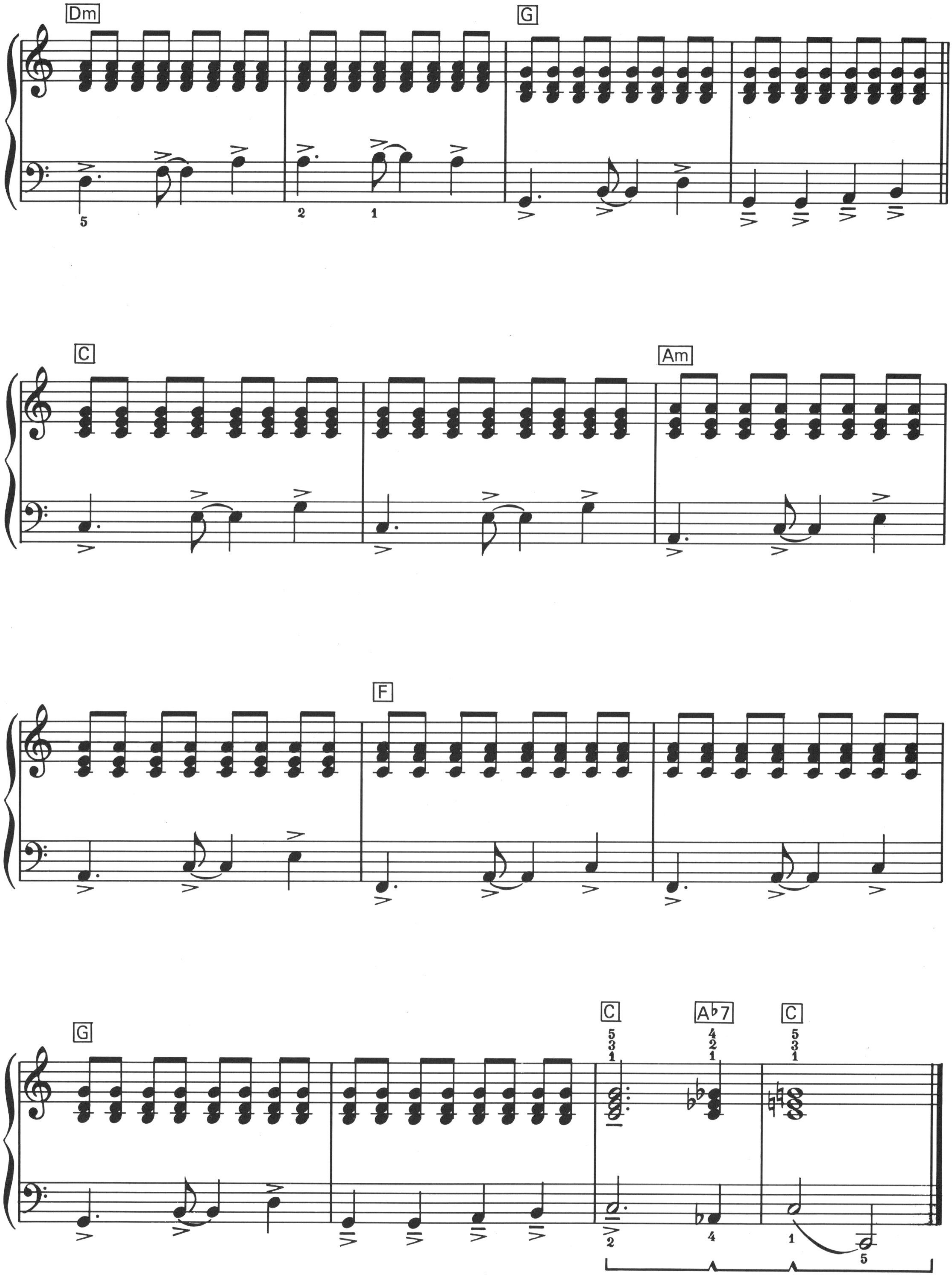
Dm
G
C
Am
F
G
C
A♭7
C

Still Loving You
(SCORPIONS)

Music and Words by Klaus Meine/Rudolf Schenker
Arr.: Hans-Günter Heumann

*Anmerkung für den Lehrer: Die internationale Schreibweise für den Ton H ist B.

© Copyright 1984 Edition Arabella Musik Muenchen, Germany.
Universal Music Publishing MGB Limited.
All rights in Germany administered by Musik Edition Discoton GmbH (a division of Universal Music Publishing Group).
All Rights Reserved. International Copyright Secured.

B. & Co. 24 928

Alle Rechte vorbehalten
All rights reserved

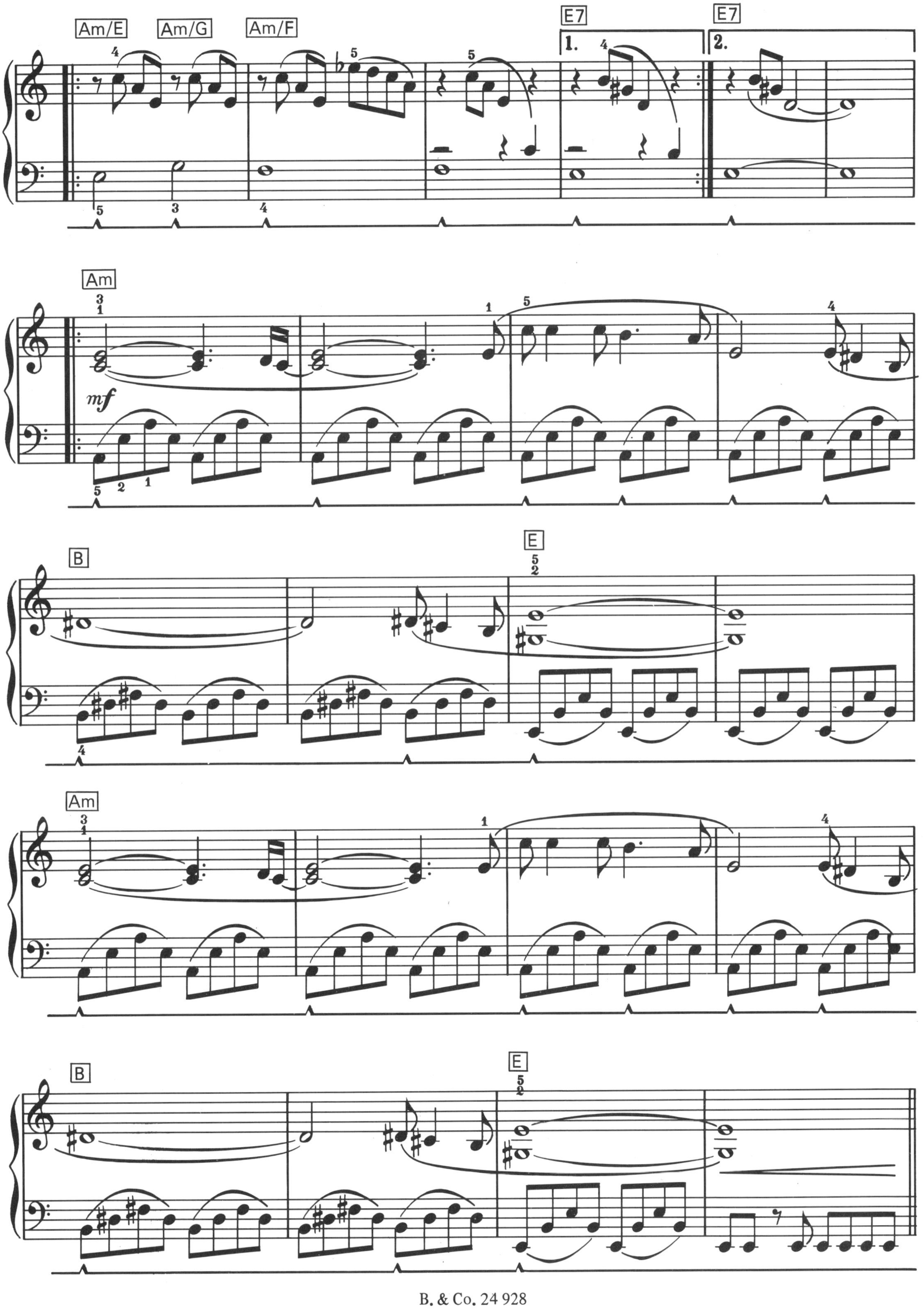
Am/E
Am/G
Am/F
E7
1.
E7
2.
Am
mf
B
E
Am
B
E

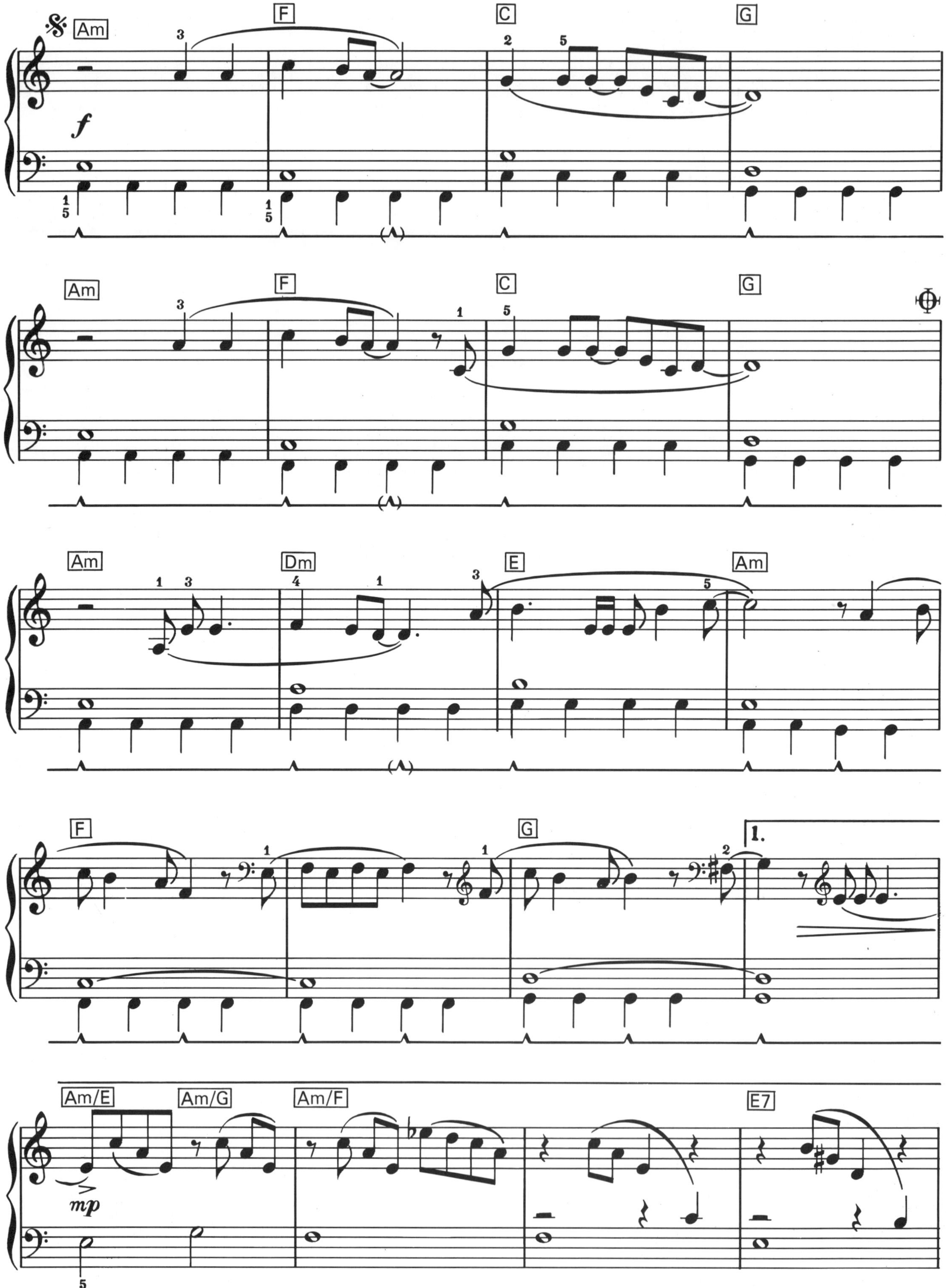
Am
F
C
G
f
Am
F
C
G
Am
Dm
E
Am
F
G
1.
Am/E
Am/G
Am/F
E7
mp

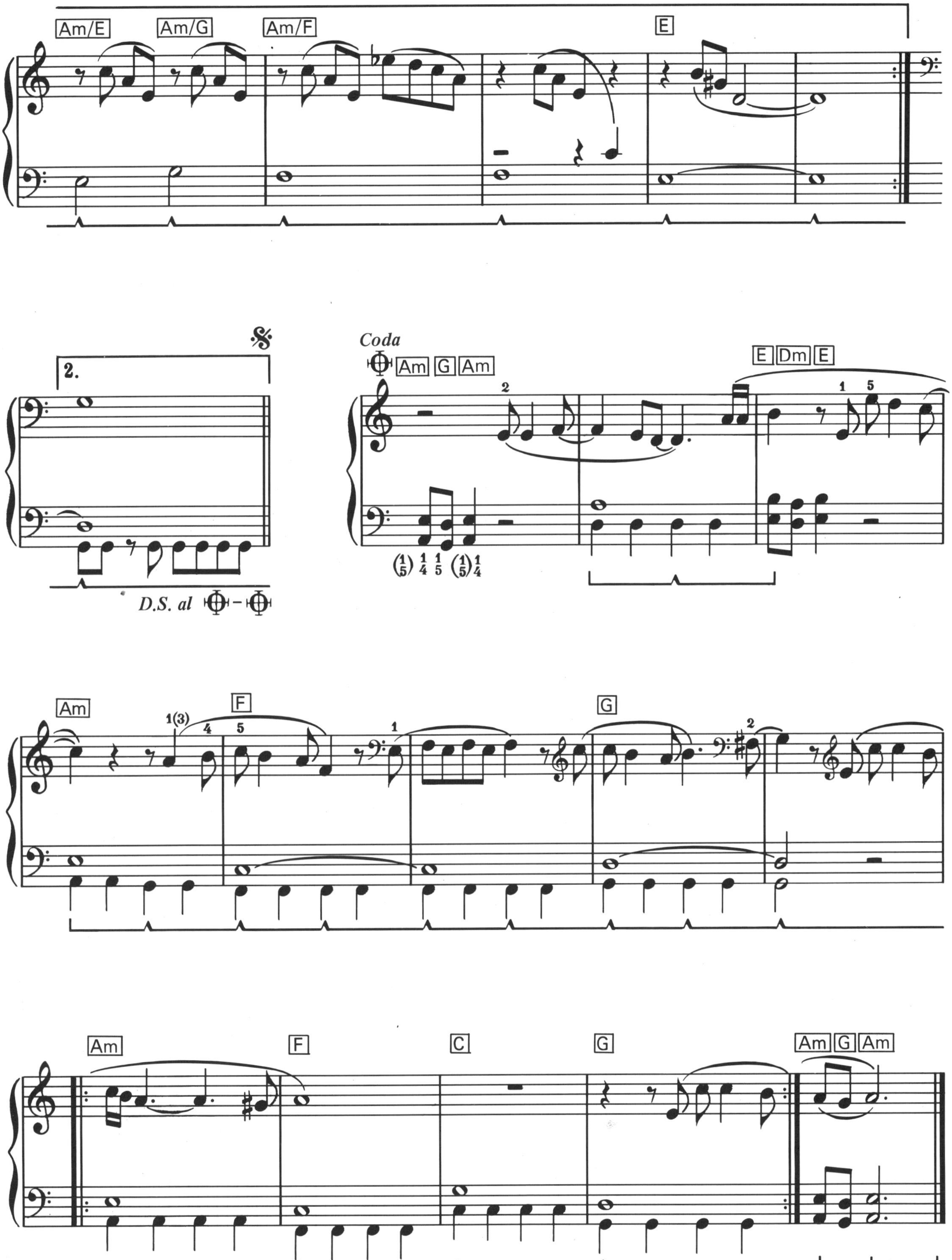
Am/E
Am/G
Am/F
E
2.
D.S. al Φ - Φ
Coda
Am G Am
E Dm E
Am
F
G
Am
F
C
G
Am G Am

That's My Sound

(Rock Ballad)

Music by Hans-Günter Heumann

Andante con espressione M.M. ♩ = 88-92

*Amerkung für den Lehrer: Die internationale Schreibweise für den Ton H ist B.

© Copyright MCMXC by BOSWORTH & CO.
BOSWORTH & CO., KÖLN-WIEN-LONDON

Alle Rechte vorbehalten
All rights reserved

B. & Co. 24 928

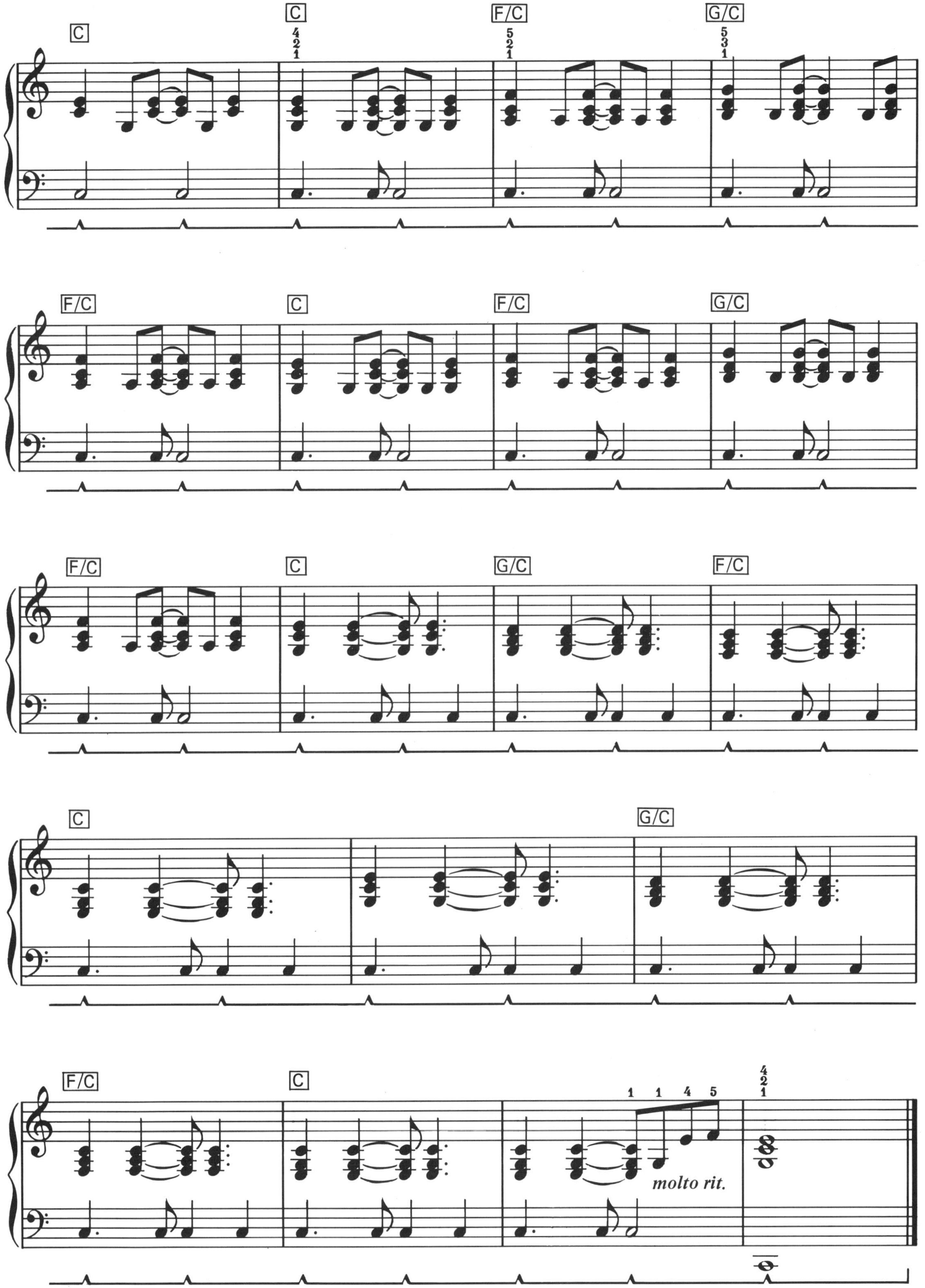
C
C
F/C
G/C
F/C
C
F/C
G/C
F/C
C
G/C
F/C
C
G/C
F/C
C
molto rit.

Oh, Pretty Woman

(ROY ORBISON)

Words and Music by Roy Orbison & Bill Dees
Arr.: Hans-Günter Heumann

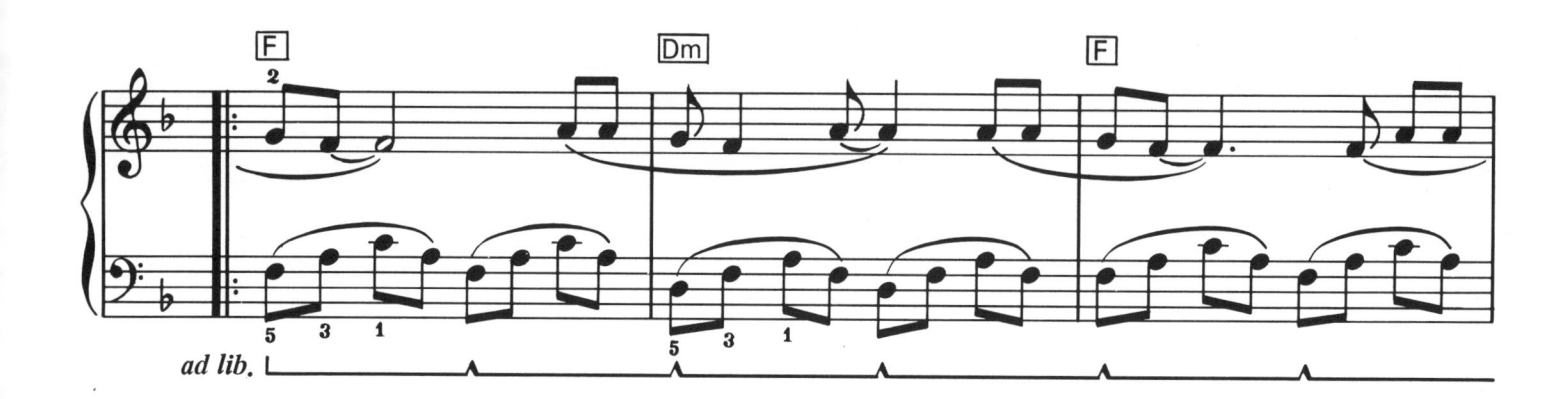

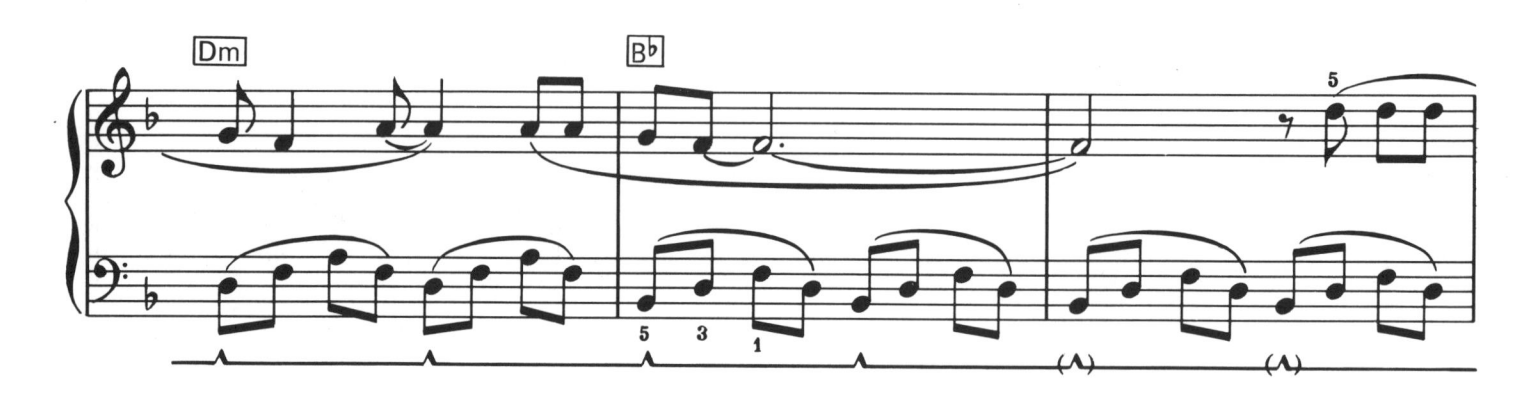

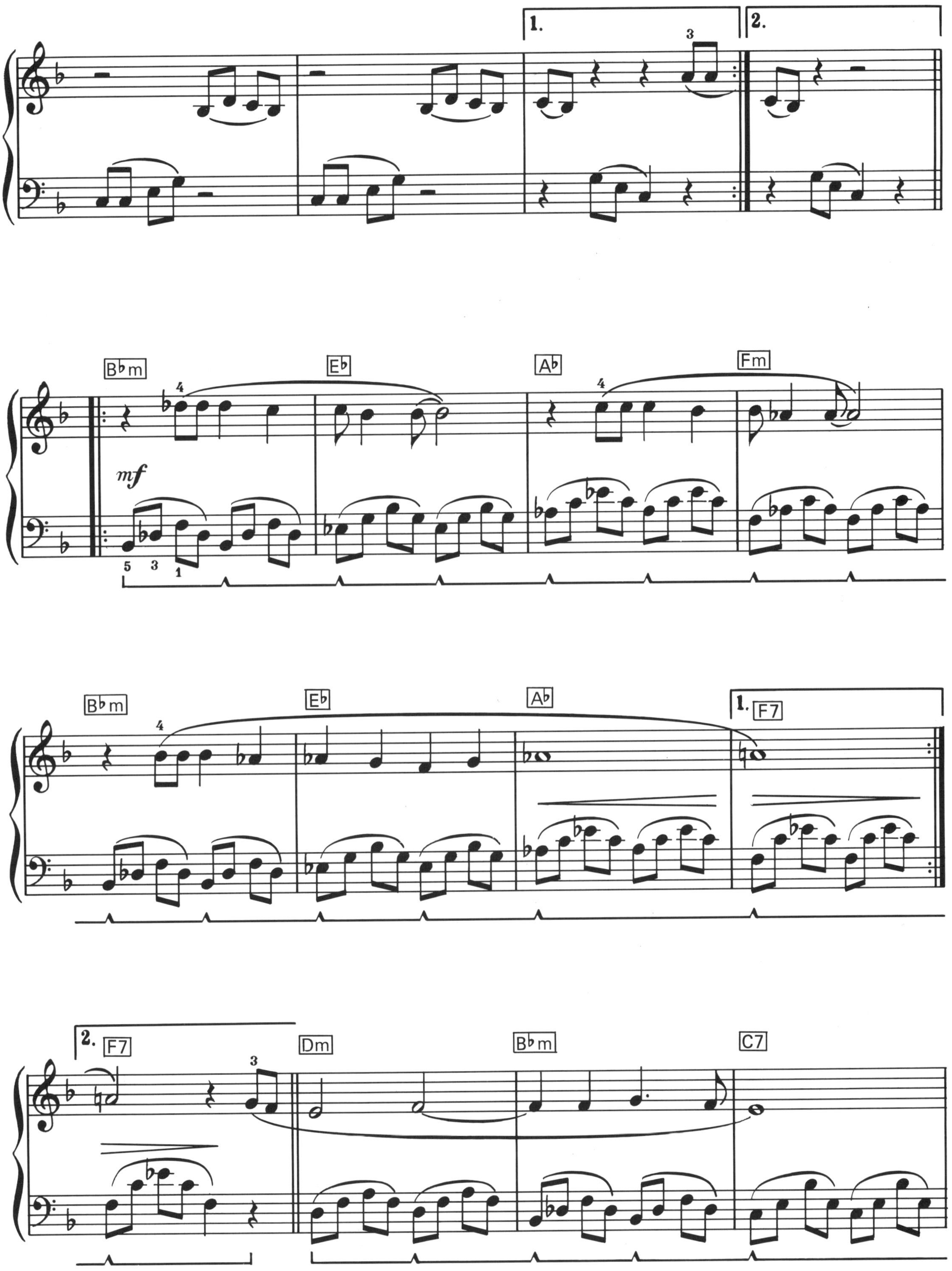
1.
2.
B♭m
E♭
A♭
Fm
mf
B♭m
E♭
A♭
1. F7
2. F7
Dm
B♭m
C7

18

Midnight Special

Traditional
Arr.: Hans-Günter Heumann

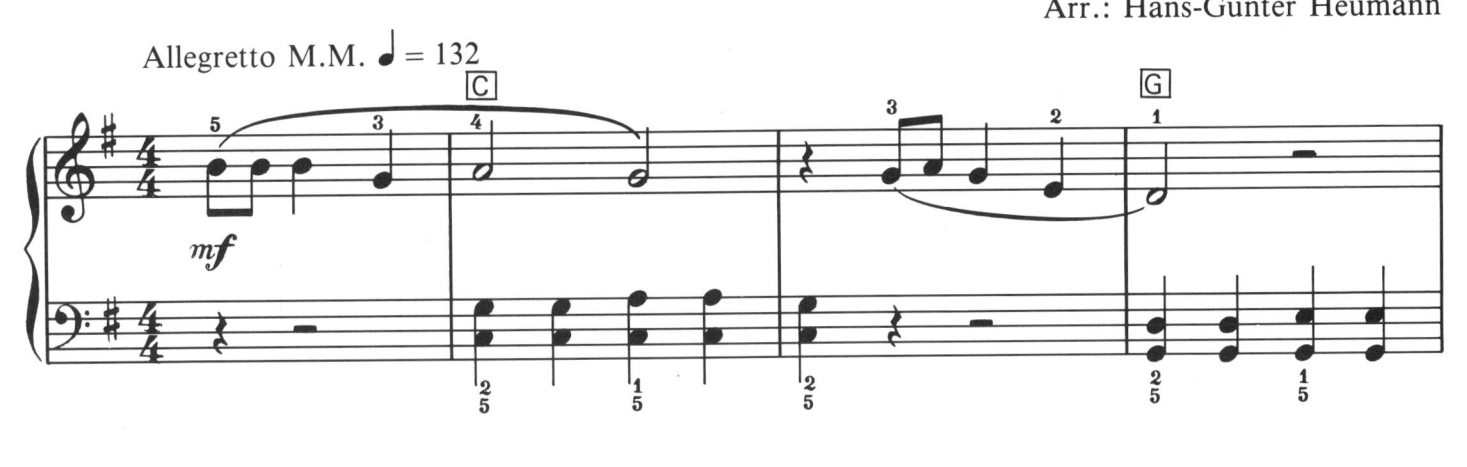

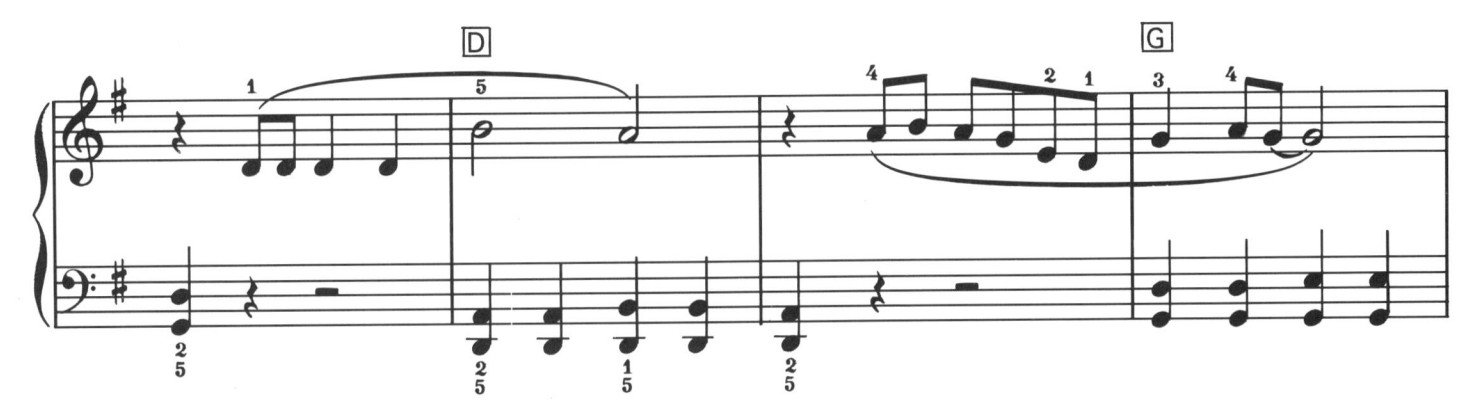

Rock My Soul

Traditional
Arr.: Hans-Günter Heumann

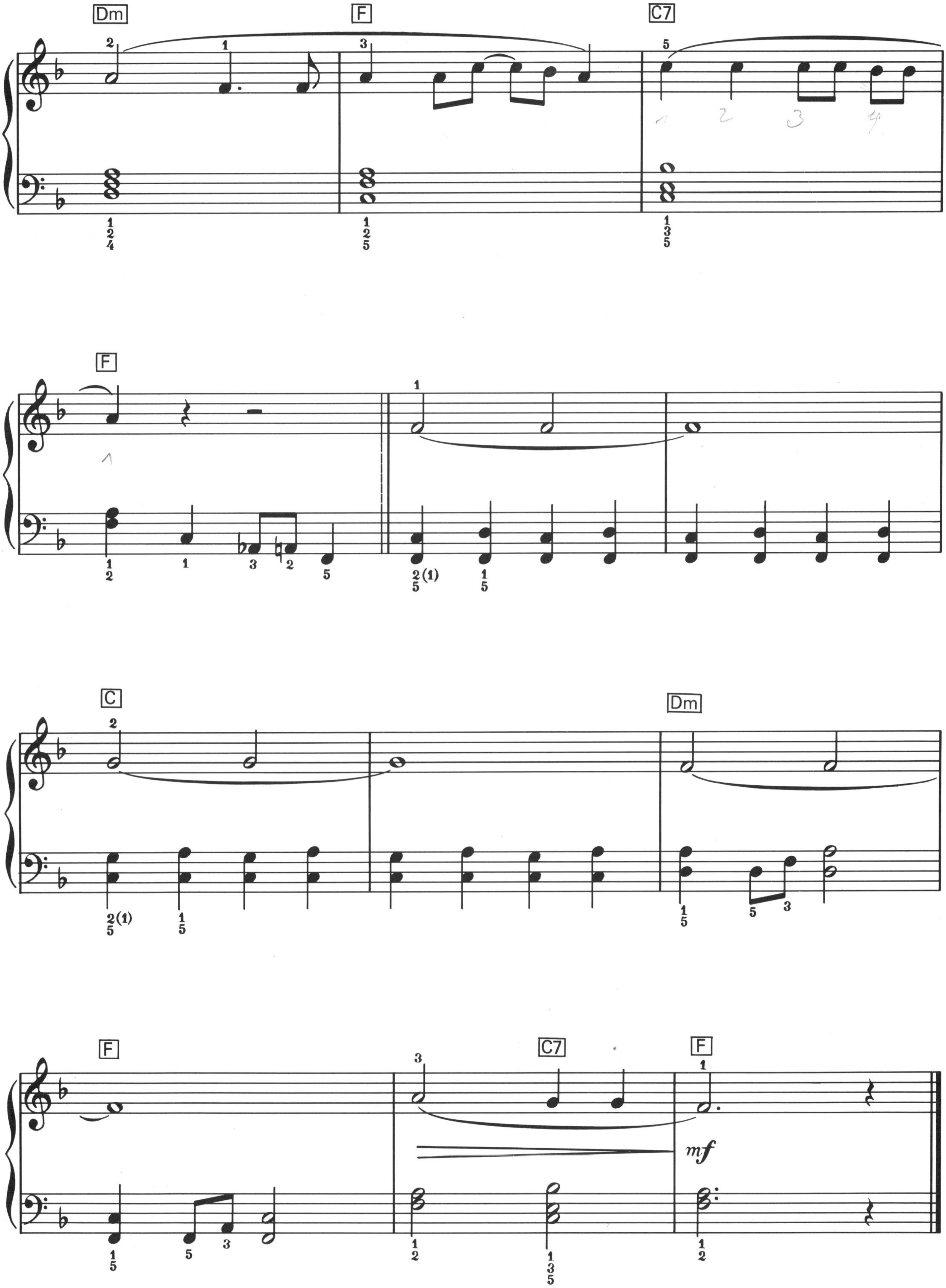
Dm
F
C7
F
C
Dm
F
C7
F
mf

Hit Boogie

Music by Hans-Günter Heumann

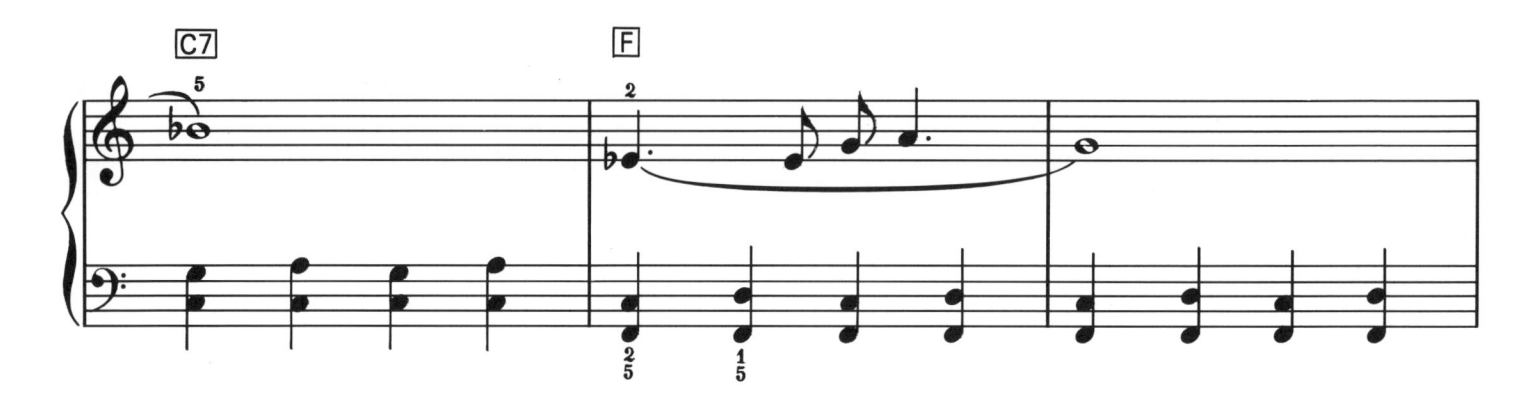

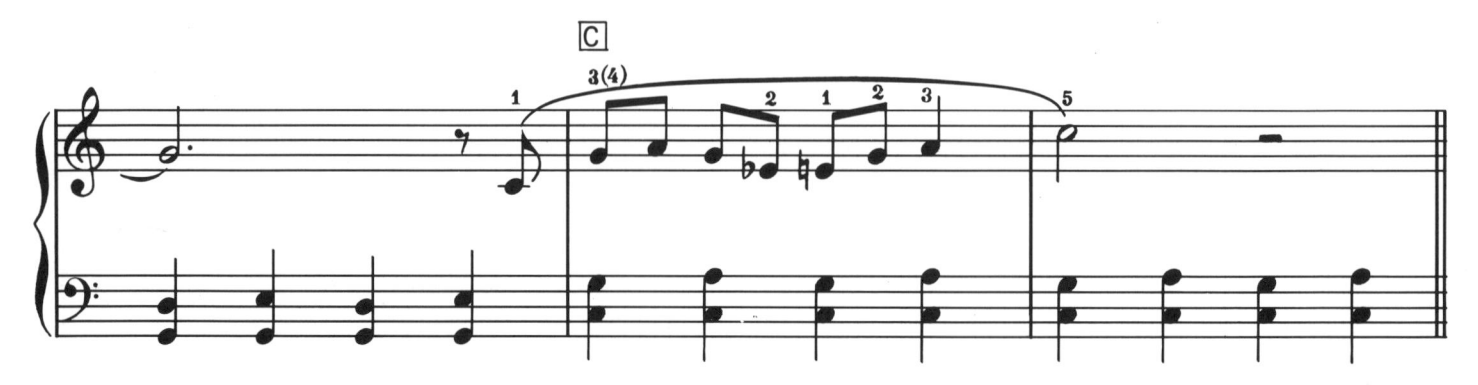

Allegretto M.M. ♩ = 120-126

23

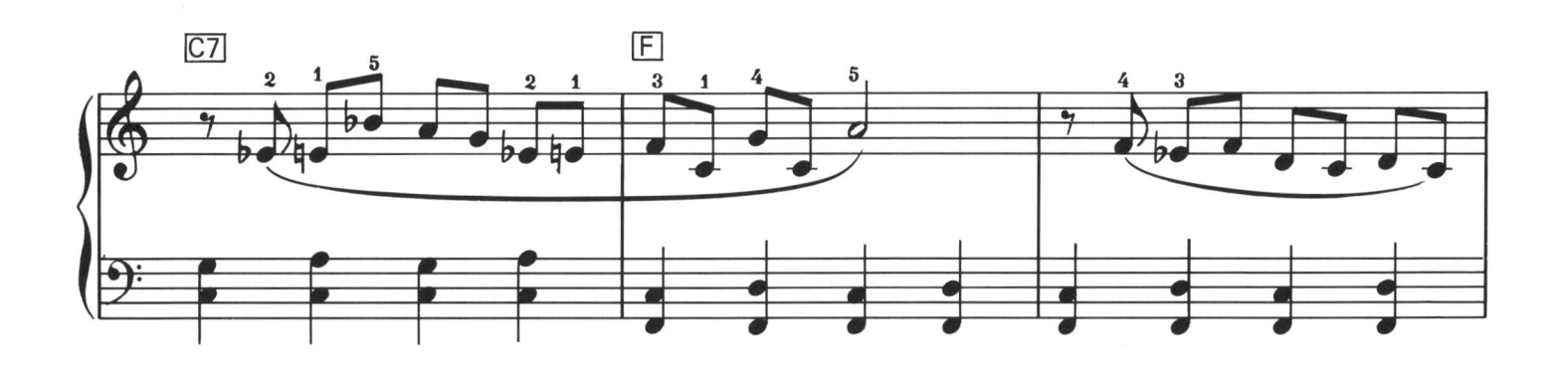

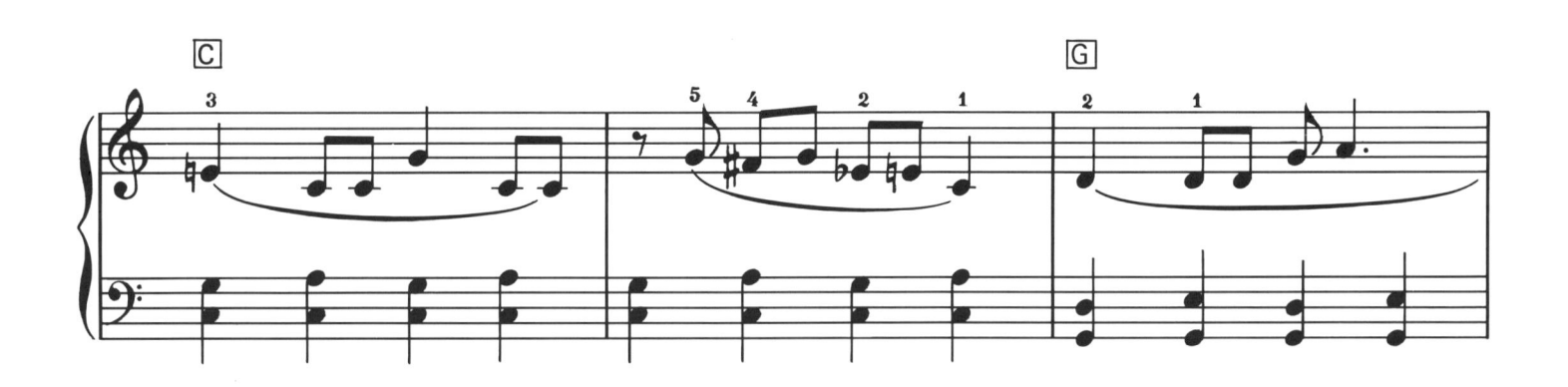

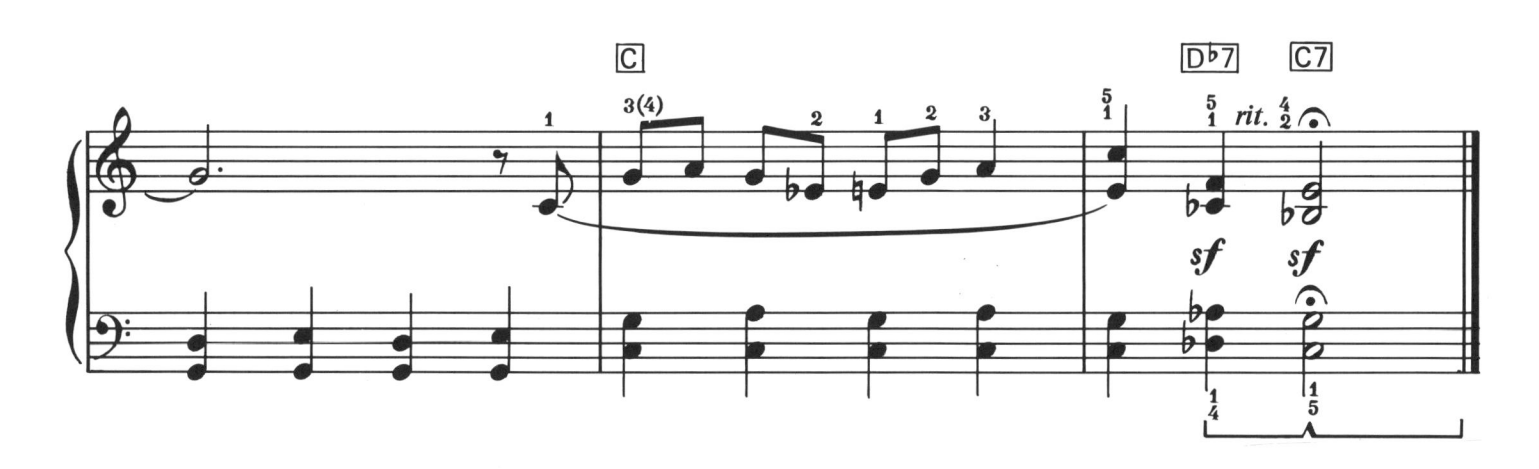

Im gleichen Verlag erschienen / Also available

HANS-GÜNTER HEUMANN
MUSIC FROM THE SHOWS

BELIEBTE MUSICAL-MELODIEN
IN LEICHTESTER FASSUNG
FÜR KLAVIER / KEYBOARD

POPULAR MELODIES FROM THE SHOWS
IN THE EASIEST POSSIBLE SETTINGS
FOR PIANO / KEYBOARD

EDELWEISS ("The Sound of Music")
TRY TO REMEMBER ("The Fantasticks")
OH, WHAT A BEAUTIFUL MORNING ("Oklahoma")
DON'T CRY FOR ME, ARGENTINA ("Evita")
CABARET ("Cabaret")
WUNDERBAR ("Kiss Me Kate")
SUMMERTIME ("Porgy and Bess")
HELLO DOLLY ("Hello Dolly")
SUNRISE, SUNSET ("Anatevka")

BoE 4036

BOSWORTH EDITION